5th Grade Math
Volume 1

© 2013 OnBoard Academics, Inc
Newburyport, MA 01950
800-596-3175
www.onboardacademics.com

ISBN: 978-1494857356

Table of Contents

Words and Numbers

Key Vocabulary

place value

million billion

What is the value of 7?

billions			millions			thousands			ones		
hundred billions	ten billions	billions	hundred millions	ten millions	millions	hundred thousands	ten thousands	thousands	hundreds	tens	ones

47,450,000 _____

3,732,400 _____

275,438 _____

674,490,000 _____

Complete these word numbers.

Can you state these facts empirically?

Number	Word Number
1,206,000	One _____ two hundred six thousand
476,500	Four hundred seventy-six _____ five hundred
5,908,000	Five _____ nine hundred eight _____
4,000,000,677	Four _____ six hundred seventy-seven

hundred   thousand   million   billion

_____

The Empire State Building is _____ inches high

In words: _____

You need to climb _____ steps to get to the top

_____

The building weighs _____ tons

_____

The total construction cost was almost $ _____

_____

1,860   41,000,000   17,488   365,000

Write the number.

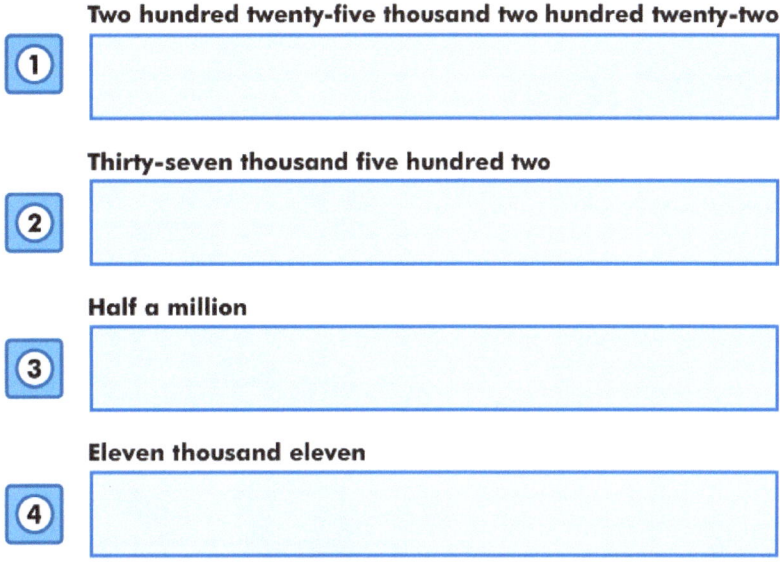

Two hundred twenty-five thousand two hundred twenty-two

1

Thirty-seven thousand five hundred two

2

Half a million

3

Eleven thousand eleven

4

Match the word and the number by drawing a line from one to the other

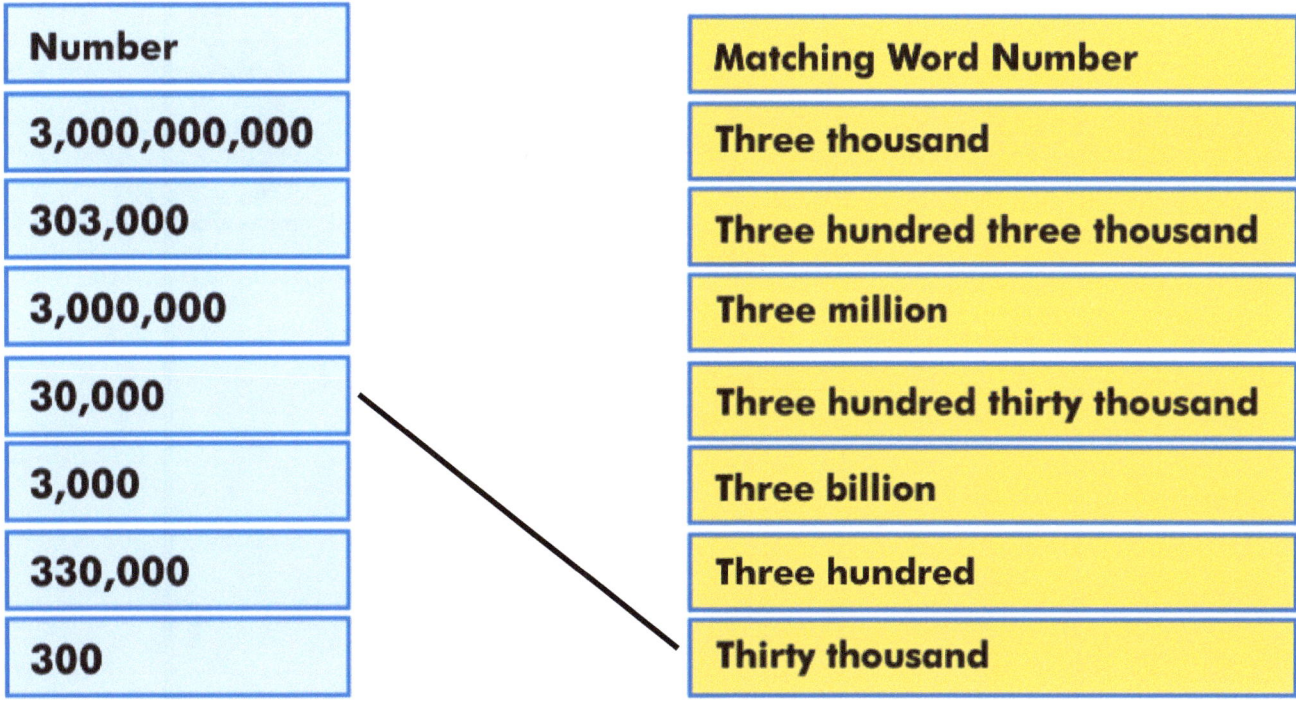

Number
3,000,000,000
303,000
3,000,000
30,000
3,000
330,000
300

Matching Word Number
Three thousand
Three hundred three thousand
Three million
Three hundred thirty thousand
Three billion
Three hundred
Thirty thousand

Order these numbers.

Greatest

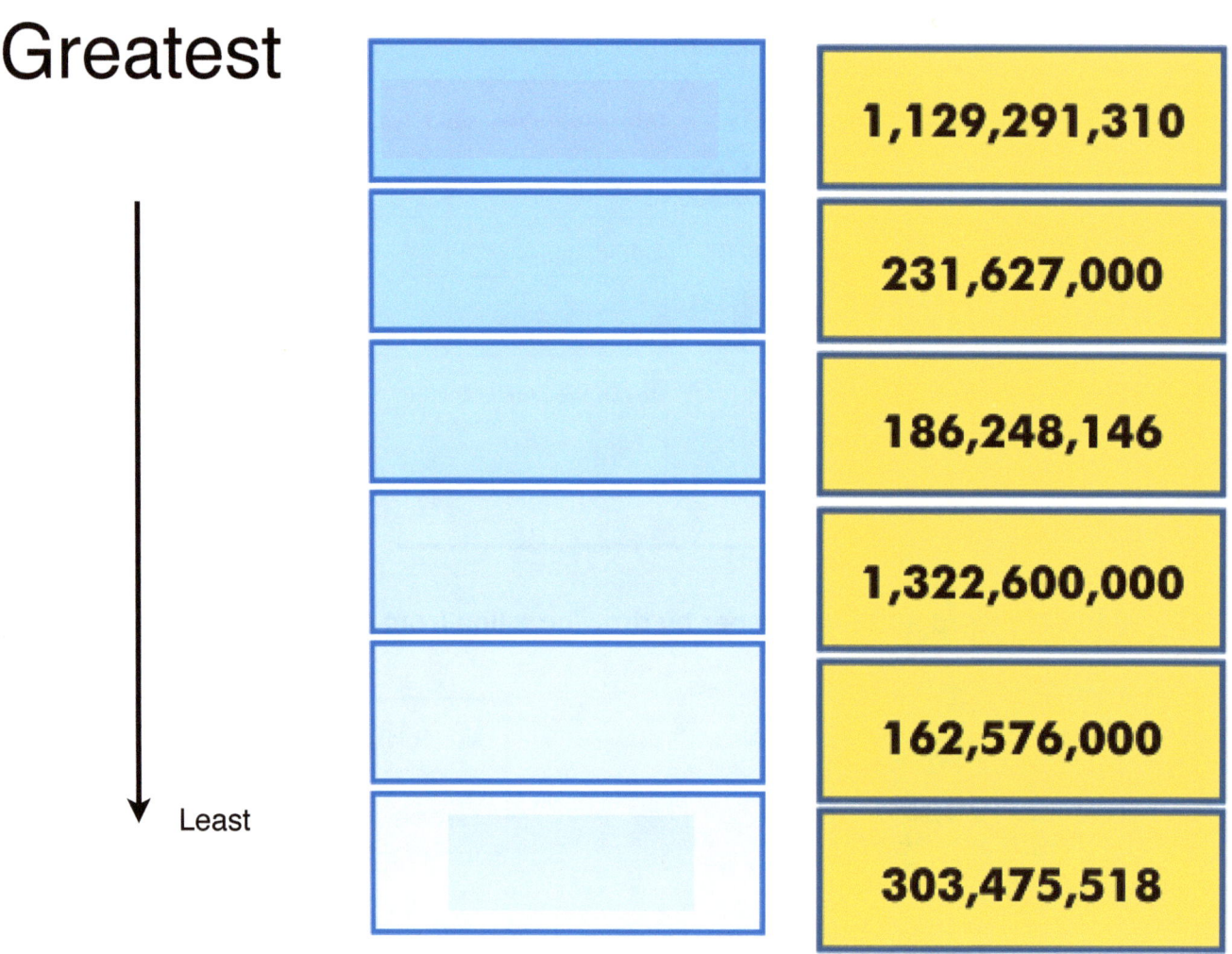

Least

1,129,291,310

231,627,000

186,248,146

1,322,600,000

162,576,000

303,475,518

Here's a stretch exercise. These populations are lined up with a country pictured on the right. Can you name the country?

1,322,600,000
1,129,291,310
303,475,518
231,627,000
186,248,146
162,576,000

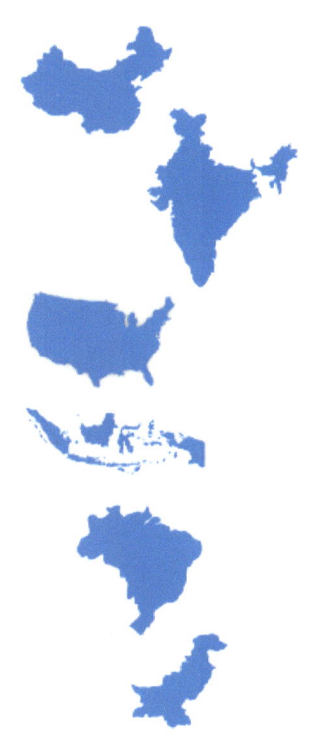

Name_____

Words and Numbers Quiz

1   True or false, ninety-nine billion is written as 99,000,000,000?

2   Which set of numbers is not ordered correctly from least to greatest?

A   4,000   40,000   40,004   40,040   40,444

B   100,000   108,909   108,999   180,999   198,999

C   8,000,000   8,999,999   9,009,888   9,008,999

D   99,999   111,111   999,999   1,111,111

3   Write this number:
nine million ninety-thousand nine hundred

4   Write this number:
eighty-seven thousand ninety-eight

Expanded Notation

Key Vocabulary

expanded notation

standard form

What is the value of the red digit?

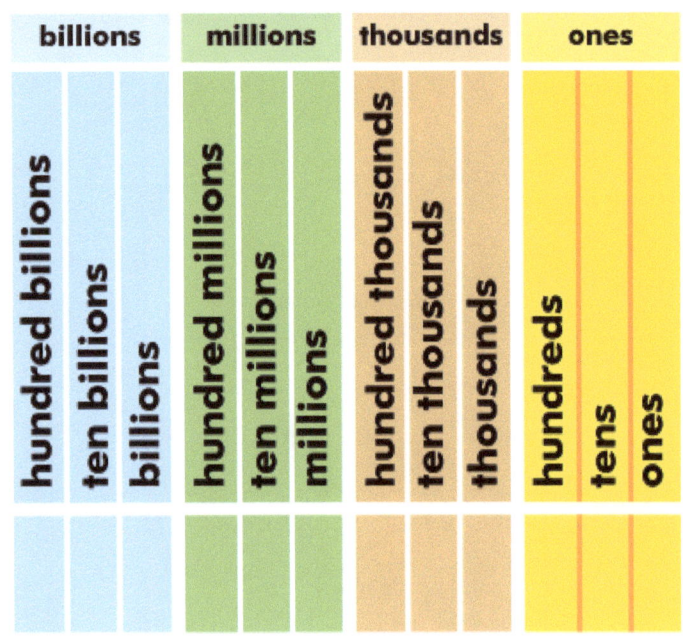

3,9̲32,400 _____

2̲65,438 _____

43̲,450,000 _____

5̲,614,490,000 _____

Standard Form and Expanded Notation

Standard form

8, 6 3 5

Expanded notation

$$\left(8 \times \boxed{1,000}\right) + \left(6 \times \boxed{100}\right) + \left(3 \times \boxed{10}\right) + \left(5 \times \boxed{1}\right)$$

Write using expanded notation.

500,050

701,301

1,010,010

Write in standard form.

$$\left(3 \times 1{,}000{,}000\right)$$
$$+ \left(4 \times 10{,}000\right)$$
$$+ \left(3 \times 1{,}000\right)$$
$$+ \left(5 \times 100\right)$$
$$+ \left(1 \times 1\right)$$

$$\left(5 \times 100{,}000{,}000\right)$$
$$+ \left(8 \times 1{,}000{,}000\right)$$
$$+ \left(4 \times 10{,}000\right)$$
$$+ \left(9 \times 1{,}000\right)$$
$$+ \left(9 \times 100\right)$$

Connect the standard form and the expanded notation.

11,011	(1 x 10,000) + (1 x 1,000) + (1 x 100)
11,100	(1 x 10,000) + (1 x 1,000) + (1 x 10)
11,001	(1 x 10,000) + (1 x 1,000) + (1 x 10) + (1 x 1)
11,010	(1 x 10,000) + (1 x 1,000) + (1 x 1)

Name_____

Expanded Notation Quiz

1 True or false? 250,000 written in expanded notation is equal to (2 x 100,000) + (5 x 1,000)

2 (5 x 10,000) + (7 x 100) + (9 x 10) + (9 x 1) = ?

- **A** 57,091
- **B** 57,099
- **C** 50,799
- **D** 50,791

3 901,004 = (9 x 100,000) + (1 x ?) + (4 x 1)

4 Write in standard form:
(1 x 1,000,000) + (1 x 10,000) + (1 x 100) + (1 x 1)

Estimation

Key Vocabulary

estimation

reasonable estimate

rounding

Sometimes you might not need an exact answer to a problem or you want to check to see if an answer is reasonable.

These situations are excellent for estimation. There are several different methods of estimation.

_____

Estimate sums and differences by rounding.

Estimate the difference in height between these skyscrapers.

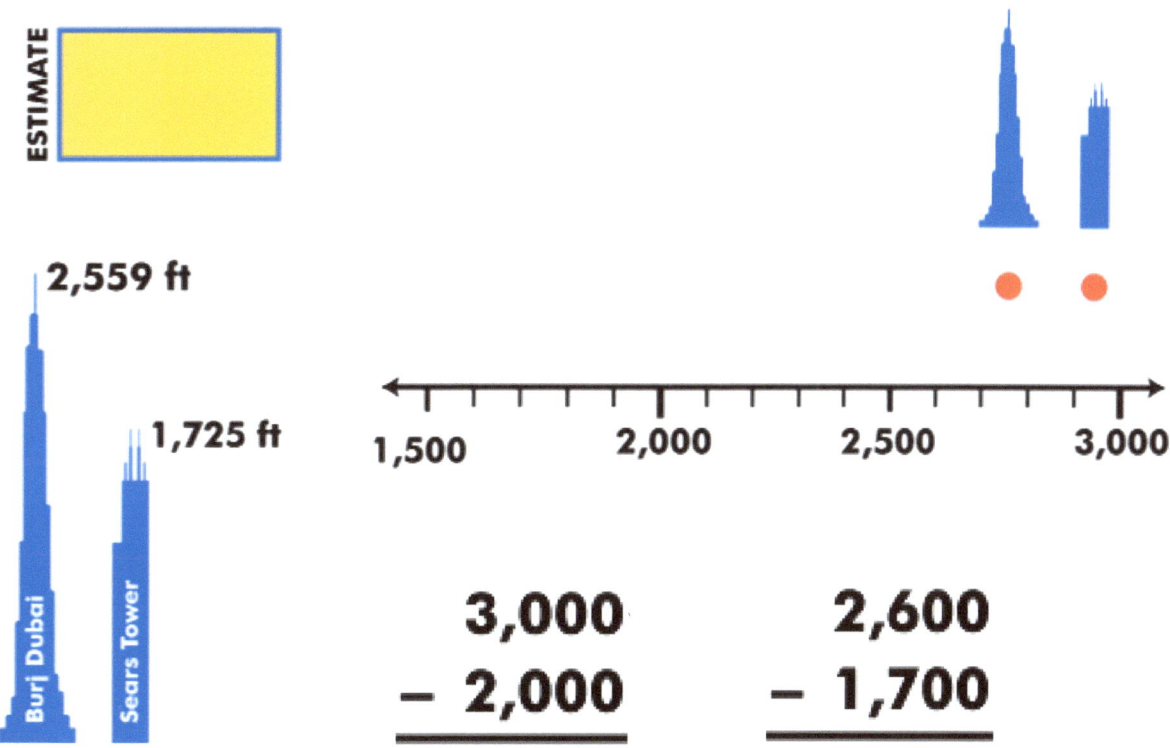

ESTIMATE

Burj Dubai — 2,559 ft

Sears Tower — 1,725 ft

1,500   2,000   2,500   3,000

$$\begin{array}{r} 3,000 \\ -\ 2,000 \\ \hline \end{array}$$

$$\begin{array}{r} 2,600 \\ -\ 1,700 \\ \hline \end{array}$$

Making a reasonable guess using a clustering strategy.
Clustering works best when all the numbers round to the same number.

Complete the clustering estimation column.

Performance	Audience	Cluster
Weds	269	300
Thurs	282	300
Fri	325	300
Sat – Matinee	273	300
Sat – Evening	334	300
TOTAL	1,483	

School Production of
MACBETH
JAN 7 — JAN 10
Free

Nancy Webber
as Lady Macbeth

Javier Morales
as Macbeth

What is the difference between the actual total and the clustering total? _____

Why is clustering a good choice for this estimation? _____

_____

Estimation using compatible numbers.
Compatible numbers are numbers that are easy to commute with mentally.
Can you round to compatible numbers to solve this equation?

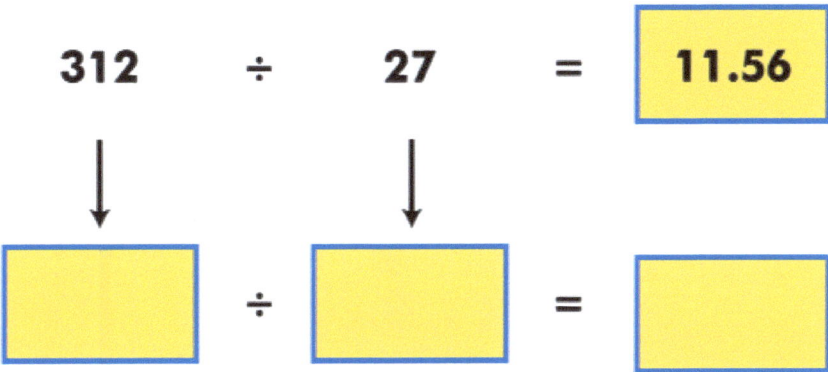

Use basic division facts to simplify the problem

Rounding
Estimate the products by rounding.
Complete the both problems, one using actual numbers and one using rounding. What is the difference between the estimation and the rounding products? _____

$$8.8 \quad \times \quad 425 \quad = \quad \boxed{}$$

$$\downarrow \qquad\qquad \downarrow$$

$$\boxed{} \quad \times \quad \boxed{} \quad = \quad \boxed{}$$

Round each factor to the greatest place

Front End Estimation

An estimation strategy that works for some addition and subtraction problems is front-end estimation. In this case you simply add or subtract the left column of digits.

Complete the front end estimation and measure it against the actual total. What is the difference? _____

TECH ★ MART	
Laptop	$ 875
40" Monitor	$ 454
Printer	$ 223
Router	$125
	_____
Total	_____
● ANSWER	

Front-End	Other Digits
800	75
400	54
	23
200	
100	25
Estimate	**Estimate**
$	$

Practice Your New Estimation Techniques
Choose the best estimate by checking the box.

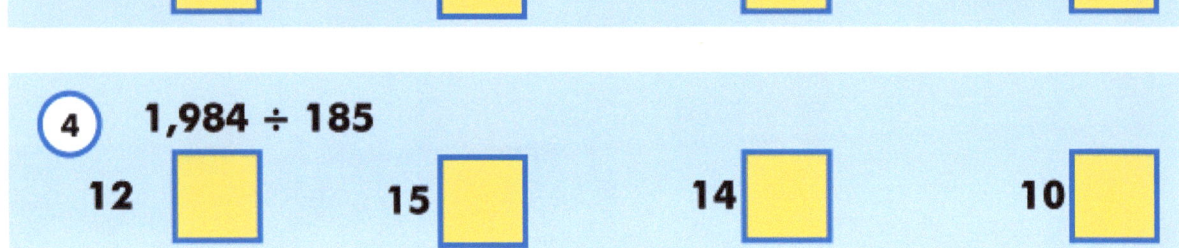

1   1,456 + 1,425 + 1,582 + 1544 + 1,472

7,500 ☐   8,000 ☐   5,000 ☐   10,000 ☐

2   25,422 + 57,833 − 9,742

91,000 ☐   73,000 ☐   75,000 ☐   80,000 ☐

3   821 + 724 + 610 + 1,205

2,500 ☐   4,000 ☐   3,300 ☐   3,000 ☐

4   1,984 ÷ 185

12 ☐   15 ☐   14 ☐   10 ☐

What estimation techniques did you use?

1. _____

2. _____

3. _____

4. _____

Name_____

Estimation Quiz

1 Choose the best estimate for 85 + 78 + 81 + 74 + 83
380   400   800   395
A   B   C   D

2 Choose the best estimate for 1.56 + 2.71 + 3.26 + 5.40
11   12   13   14
A   B   C   D

3 287 students are going on a camping trip. If their tents sleep 7, estimate how many tents will be needed.
60   30   50   40
A   B   C   D

4 Each student pays $56 for the trip. Estimate how much money is paid in total by the 287 students?
$1,600 $18,000 $8,000 $24,000
A   B   C   D

Add & Subtract Whole Numbers

Key Vocabulary

regroup

strategy

Study the problem below to review adding whole numbers.

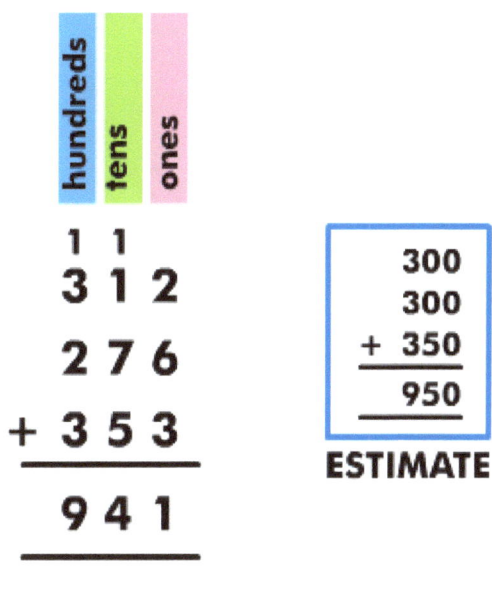

Study the problem below to review subtracting whole numbers.

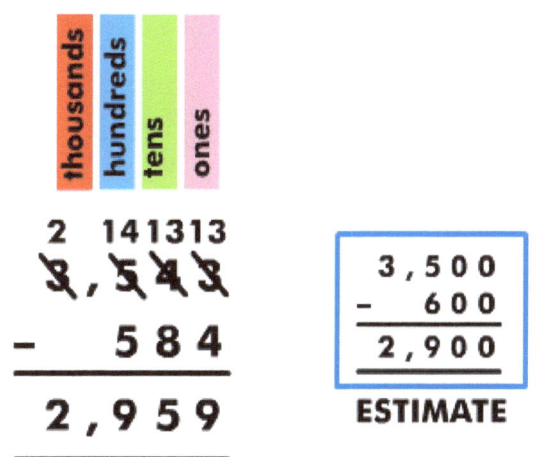

Practice estimating and adding whole numbers.

```
   1,123        10,767         4,135
 +   445       + 9,649        16,237
 _____      _____      + 14,494
                             _____
 _____      _____      _____
```

ESTIMATE   ESTIMATE   ESTIMATE

Practice estimating and subtracting whole numbers (one has a twist).

```
   2,741        11,747        16,237
 –   608       – 9,649       – 4,494
 _____      _____      _____

 _____      _____      _____
```

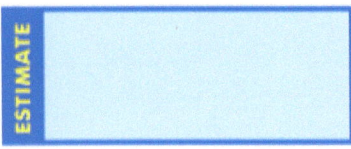  

ESTIMATE   ESTIMATE   ESTIMATE

Mrs. Jones flew form Miami to San Francisco via Atlanta. How many extra miles was this versus a direct flight?_____

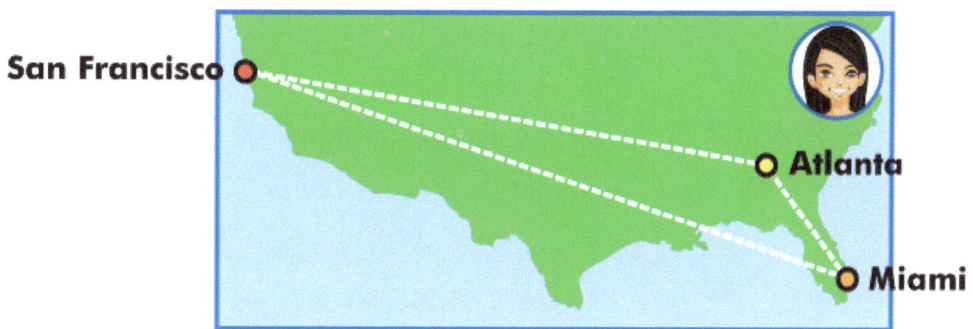

Miami to San Francisco 2,601 miles

Miami to Atlanta 608 miles

Atlanta to San Francisco 2,133 miles

Mrs. Jones got a direct flight back. What was her total round trip?

Name_____

Add & Subtract Whole Numbers Quiz

1 True or false? The total population of the countries in Table 1 is 443,230,979.

2 How much larger is the population of the USA than the population of Canada?

(A) 334,530,088

(B) 67,749,806

(C) 276, 749,806

(D) 267,749,806

Table 1

Country	Population
USA	301,139,947
Canada	33,390,141
Mexico	108,700,891

3 900 − 452 = ?

4 777 + 888 = ?

Newburyport, MA 01950

1-800-596-3175

OnBoard Academics employs teachers to make lessons for teachers! We create and publish a wide range of aligned lessons in math, science and ELA for use on most EdTech devices including whiteboard, tablets, computers and pdfs for printing.

All of our lessons are aligned to the common core, the Next Generation Science Standards and all state standards.

If you like our products please visit our website for information on individual lessons, teachers licenses, building licenses, district licenses and subscriptions.

Thank you for using OnBoard Academic products.